BIRTHDAY BOOK

HONORING

STEVEN FOX

APRIL 16, 1986

PRESENTED BY

MR. AND MRS. NATHAN FOX

SUN POWER

SUN POWER

The story of solar energy

Madeleine Yates

line illustrations by
J. S. Laughbaum

Abingdon
Nashville

PROPERTY OF
LAKEHILL PREPARATORY
SCHOOL LIBRARY

Sun Power: The Story of Solar Energy

Library of Congress Cataloging in Publication Data

YATES, MADELEINE, 1937-
Sun power.
Bibliography: p.
Summary: Defines solar energy and explains where and how it originates, its uses, how sunlight becomes electricity, and the future of this energy source.
1. Solar energy—Juvenile literature. [1. Solar energy] I. Title.
TJ810.Y37 333.8'8 81–14957 AACR2

ISBN 0-687-40627-7

Acknowledgments

Wonderful people deserve appreciation, especially those who made this book possible. Fred Tomey deserves a special tribute for opening wide the doors at JPL/NASA and providing access to information and the right contacts. Thanks to science expert Ben Casados who read my manuscript and checked all facts for scientific accuracy. Both Ben and his longtime friend, Norm Avery of the Solar Energy Research Institute, generously furnished pictures for this book. Most of all, thanks to you, dear reader, for sharing my fascination with the sun, a subject that touches all of us.

Manufactured in the United States of America

To the sunshine of my life:
Leon, Michael, Elizabeth, and Gregory Nicholas

Contents

1

WHY SOLAR ENERGY?

Look around! Wherever you go, you can see how the energy crisis has changed our lifestyle. Air conditioners are often left idle on hot days. People are using extra blankets rather than raising the heat. Gone are the days of big cars. Smaller cars that use less gas are here to stay.

Prices are soaring, partly because of the energy

crisis. It costs more to live because we need precious energy to manufacture and deliver our many products.

Despite the advertising campaigns, despite many efforts toward conservation, we still use more energy than we should. Oil provided half of the world's energy needs for 1980. Oil and gas have kept the world economy strong. But our oil and gas supplies have peaked. Our reserves are running short!

World leaders are seeking a solution to our energy shortages. Not long ago, the Jet Propulsion Laboratory presented the Department of Energy with a large lit-up panel, illustrating our energy resources for the future. If you looked at the pictures, you would notice something missing: *oil and gas!* They are the main power resources for today, but not for the future.

Unless new energy sources are developed, the time may come when you flip on the switch and no light comes on. Energy shortages may force the power companies to plan "rolling brown-outs." That is, the electricity will be turned OFF in different areas at preplanned times. Residents will be warned ahead of time, but what a terrible inconvenience!

The question is—how can we prevent this from happening? Conservation is one answer. If each of us became less wasteful, it would make a huge difference.

New forms of energy must be developed—and soon. Our scientists are exploring many challenging possibilities. Some possible energy sources are: *Solar* energy from the sun. *Geothermal* energy from underground heat. *Biomass* energy from wood, plant waste, and garbage. *Ocean thermal* energy from temperature differences in the ocean. *Wind* energy from moving air. *Tidal* energy from ocean tides. *Nuclear* energy from splitting atoms.

Among these alternatives, the sun offers the most spectacular potential. Some think that solar energy,

no matter how successful, will serve only part of our needs. Others foresee a solar-powered society in the decades ahead. They believe that we are entering the Solar Age, and that the sun will be a major energy source.

Solar energy offers many benefits. It is clean. It will not cause smog, pollute drinking water, contaminate ocean or river water, or leave dirty by-products. The sun's supply is almost endless. It has been shining for billions of years, and it will continue to shine for five billion years to come. Unlike nuclear energy, sun power leaves no radioactive waste. Bombs cannot be made from sunshine, nor can nations try to take control of the sun.

The amount of energy given off by the sun is so enormous that it is hard to imagine. Radiant energy is constantly leaving the sun's surface and drenching our earth. Only a small portion of the sun's energy penetrates our atmosphere, however. Most of it is lost in outer space.

Our lives depend on sunshine. Many forms of life and energy come from the sun.

The sun gives us food. We need food to survive. Plant life makes up our basic food. The sun shines on plants, which change the sunshine's energy to food. Plants live on sunshine and the minerals in the soil.

The sun gives us fresh air. Each day, when the sun's rays fall on leaves, life-giving oxygen is

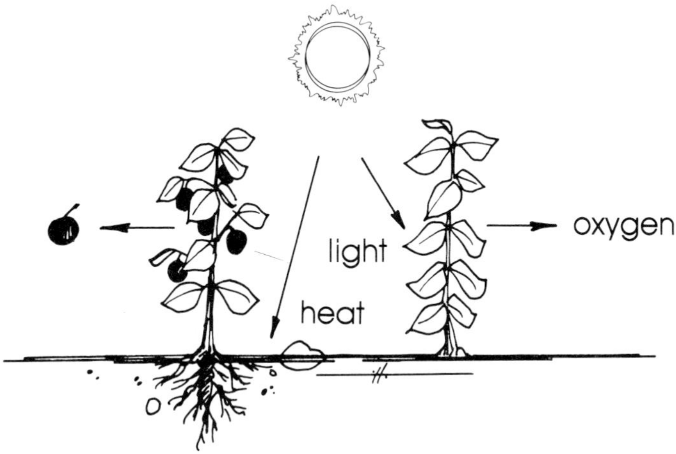

released by the plants to freshen the air we breathe.

The sun gives us daylight and moonlight. The rising sun brings day. Moonlight is really sunlight reflected off the moon's surface.

The sun gives us wind. When the sun warms the earth unevenly, the air over hot areas rises, and the

heavier cool air flows in to take its place, causing the air's constant movement.

The sun gives us rain. Its heat causes ocean water to evaporate and then fall again as rain. During this process, salt and impurities are left on the ground or in the ocean. The water becomes pure and drinkable.

The sun gives us heat. Without the sun, our earth would be frozen. Its everlasting heat and energy keep us alive.

The sun gives us fossil fuels. Oil, gas, and coal are products of the sun, too. They are decomposed remains of prehistoric plants and animals that soaked up the sun's energy millions of years ago. When we burn these fuels, we are really burning stored-up sunlight from long ago.

2

WHAT MAKES THE SUN SHINE?

For thousands of years, people have been trying to discover the truth about the sun. Aristotle recorded the sun's patterns and proclaimed that the sun's surface was smooth. One scientist felt certain that people lived *inside* the sun. Another said the

sun was a "heat engine" with a uniform temperature throughout. But they were wrong.

Our real understanding of the sun started only three hundred years ago. But the most exciting facts have been discovered within the last thirty years. Powerful telescopes and spacecraft that act as orbiting laboratories have helped us learn much about the mysterious sun.

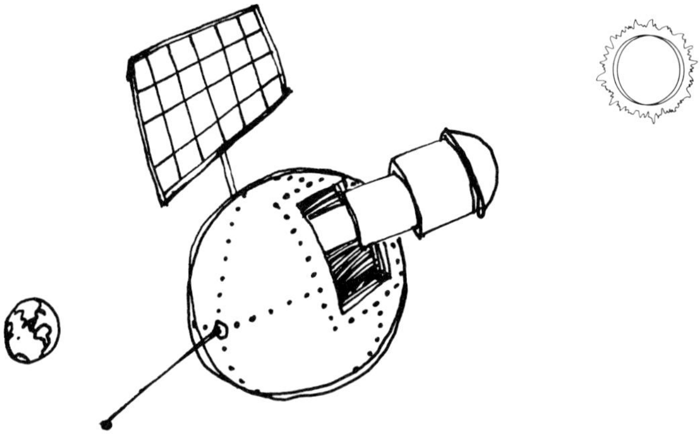

We have discovered that the sun is a typical red-hot star, like billions of others in the Milky Way galaxy. In Latin, *sol* means sun. Solar energy is energy from the sun. In one second, the sun gives off more energy than we have used since the beginning of civilization.

Earth receives only a tiny portion of the sun's total power. If we could harness our daily sunlight, it would satisfy our energy needs forever.

Energy from the sun travels at the speed of light: 186,000 miles per second. It only takes sunlight eight minutes to soar through 93 million miles of space before reaching earth.

The sun's rays are electromagnetic, similar to radio or television waves. The different wave lengths result in light we can see and heat we can feel. Although sunshine consists of a variety of radiating waves, we are most familiar with three.

Visible light is the sunlight you see. *Invisible heat waves* are the sun's heat you feel. *Ultraviolet radiation* gives you a sunburn, though you cannot see it.

Our planet is close enough to enjoy the sun's benefits, and far enough away to be safe from its tremendous heat. If the earth were closer to the sun, its radiation would not only boil all our oceans away in seconds, it would melt the entire earth.

What causes this tremendous heat? Burning fire?

If newspaper were burned, the fire would last as long as the newspaper was in flames. When the paper turned to ashes, the ash would eventually grow cold. Scientists know that the sun does not burn like newspaper, for it would have burned itself out and grown cold.

The sun's heat and energy are caused by explosions that occur each time heavy hydrogen combines to form helium. Deep inside the core, hydrogen atoms are constantly joining and changing into helium. This complex chemistry is called a *thermonuclear reaction*.

You might think of the sun as a huge hydrogen bomb in the sky. It is continually exploding. When you imagine an explosion, you visualize things blowing apart. But the sun does not blow apart.

Gravity keeps the sun together and balances the forces within. Through the force of gravity, the size of the sun remains unchanged.

The sun's gravity is twenty-eight times stronger than earth's. Since gravity affects one's weight, a human being would weigh about two tons on the sun.

Comparing our earth to the massive sun is like comparing a pinhead to a planet. While the earth has a diameter of 8,000 miles, the sun has a diameter of 864,000 miles. The sun could hold 332,000 earths within its sphere.

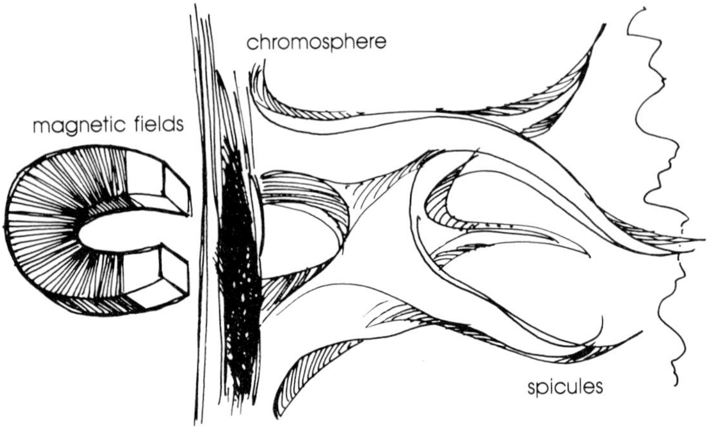

The light that shines around the sun resembles a cloudlike halo. It is called the *corona,* which means *crown.* Superheated gases of the corona are constantly boiling off, extending a million miles into space. This creates solar winds, through which heat, light, and atomic particles travel.

Closer to the sun's surface lies a heavier layer called the *chromosphere.* When the sun is more "active" than usual, the light of the chromosphere radiates outward in long spikes, or *spicules.*

The corona and chromosphere form the sun's atmosphere. They are like the thin skin on an onion when compared to the huge body of the sun.

The sun's surface begins with the *photosphere,*

or "body of light." The photosphere is densest at its deepest point—close to the sun's surface. Its thick gases bubble like hot molasses. Hotter gases from below boil up and keep the photosphere in constant motion.

Hundreds of hot spots are scattered about the sun. They surge like bright torches, winking on and off. Releases of energy cause fiery eruptions in the form of flares, plumes, and fountains. Twisted sheets of gas arch all over the sun's surface like red streamers. Sometimes these flares reach a fantastic height of 400,000 miles. One flare releases a

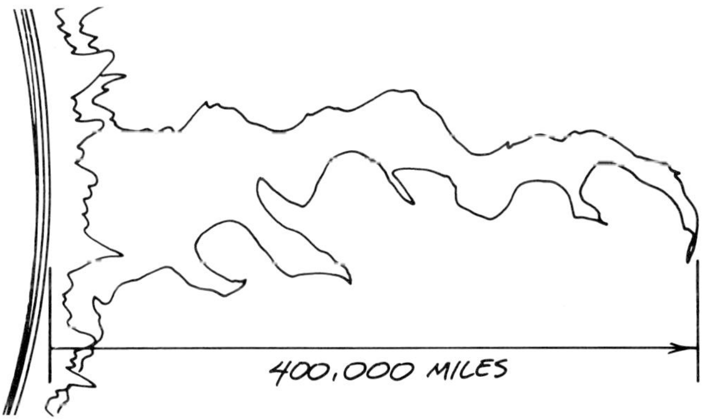

million times more energy than the biggest bomb or earthquake. Day by day, the sun's surface changes.

There are magnetic fields on the sun. They pull against the sun's gases that spiral upward, and trap them. The tangled magnetic fields and gases pull against one another. Sometimes, when a single magnetic field ties itself into a "knot," it becomes a *sunspot.* Sunspots appear darker and cooler than the surrounding surface.

Nearly all the sun's energy and light comes from the *core.* Within the sun's center, temperatures reach eighteen million degrees Fahrenheit. Core

gases are ten times denser than steel. Incredible pressures mount as overlying masses of gas press inward, undergoing steady explosions as hydrogen fuses into helium. Shock waves carry the sun's energy outward through the sun's body into outer space. As the sun rotates, its energy spins off the same way water spins from a lawn sprinkler.

An amazing feature about the sun is that it "breathes" like a giant balloon. In-and-out movements of expansion and contraction were discov-

ered by a team of French astronomers. But no one knows exactly why the sun "breathes." More is understood about the sun's atmosphere and surface than about its interior. The more we learn about the sun, the more mysterious it becomes.

3

WHEN DID SOLAR HISTORY BEGIN?

Ancient civilizations worshiped the sun as a god for thousands of years. They thought their god was surrounded by blinding light. When the sun went into an eclipse, they became fearful that their god was angry.

Early Egyptians called their sun-god Ra. Greeks and Romans believed that their sun-god drove a chariot across the sky each day. In Mexico, the Aztec Indians offered human sacrifices to show devotion to Huitzilopochtli. The Natchez Indians of North America burned perpetual fires to honor their god, and they called their priest-ruler "Great Sun." Of all the sun-god temples, the most dazzling was built in Peru by the Incas, who covered interior altars and exterior walls with pure gold.

Around 2,500 years ago, the Greeks suffered a fuel shortage. Forests were stripped bare because the Greeks were using wood for everything—fuel, heating, cooking, and building homes and ships.

Socrates, a Greek philosopher, told people to build their homes higher on the south side. By doing this, they could take advantage of the sun's path in the sky. Those who followed his advice found their homes were definitely cooler in summer and warmer in winter. Homes in Delos and

Olynthus used solar heat. The entire city of Priene was built on the southern slopes of Mt. Mycale to capture more sun. Priene, in Greece, was probably the first solar community in the world.

The Romans forged ahead with solar architecture. They used glass windows to trap the sun's heat. Pliny, a Roman writer, often bragged about how his villa collected and increased the warmth of the sun.

A law stated that anyone who blocked the sunlight from his neighbor's house would face a lawsuit. Romans viewed sunlight as a resource. Citizens had the right to receive full light from the sky directly above their property. Anyone who blocked that light was breaking the law.

Rome was famous for its public baths. Eight hundred were scattered across the city. The largest bath held two thousand bathers. Some of the baths were quite lavish. Outside, there were gardens, statues, and fancy walkways. Inside, there were facilities for cold baths, warm baths, and steam baths. Water was heated by wood-burning furnaces connected to a network of steam pipes beneath the floor. The Romans also heated their water with sunlight shining through glass windows in the ceiling.

Seventh-century Indians living in Arizona were cliff-dwellers. They built their homes in canyon walls. One famous cliff dwelling was Montezuma's Castle. The castle was heated and cooled by the sun, and the Indians derived seasonal comfort without air conditioning or natural gas. On winter days, intense sunlight beat against the outer walls. The canyon walls absorbed the sun's heat like a sponge. When the temperature dropped at night, the stored-up heat radiated outward. Thus, the walls acted as a radiator and heated the rooms on cold nights.

In summer, a rock overhang shaded the dwelling from the summer sun. The Indians used the same principles then, that our energy experts are now using in architectural design.

During the Renaissance (in the 1400s), greenhouses became popular. They were used to grow oranges and lemons year round. The greenhouses trapped and magnified so much solar heat that citrus plants grew in winter.

Solar hot boxes were widely used in Europe in the eighteenth century. Made of glass, they were used to absorb the sun's heat for cooking. Later, a better box was designed. It resembled a large window of glass and slate. This was the beginning of our solar panels. Massachusetts Institute of Technology (M.I.T.) spent twenty years researching solar panels. During this time M.I.T. learned how the panel should be constructed, how to tilt it to capture the most heat, and the most effective kinds of materials.

In 1901, an ostrich farm in Pasadena was

publicized because its water pump used a 3.5-kilowatt solar engine. The generator was built by one of the first solar companies in America, the Solar Water Company of Boston.

Today, many people are using the sun's power as an alternative to conventional energy sources. An example is a house in Utah that uses a passive (natural) solar design to trap the sun's heat. There are no rooftop collectors. The house is an "earth berm," which means that it is actually built into the mountain. Sunlight shines through huge south-facing windows. Dark stone walls retain the heat, which spreads from room to room. The house has a backup system of electric heat pumps for days when the sun does not shine. In summer no air conditioning is needed because the house circulates the cool mountain air.

4

WHO PIONEERED SUN RESEARCH?

Solar pioneers experimented with sunlight. In different ways they put sunlight to practical use. Through the centuries their efforts have expanded our understanding of sunlight and its many applications.

One timeless legend tells about Archimedes, who lived in ancient Greece. An enemy fleet of Roman ships was approaching Syracuse. The king had to do something quickly.

Rather than turning to his military advisers, the king rushed out to find Archimedes, an absent-minded mathematician. He found him in the marketplace, sitting on a big rock and drawing math equations in the sand.

"What should we do?" cried the king.

Archimedes sighed. The king was always coming to him with problems. All he wanted in life was to work with mathematics. The sooner he defeated the enemy ships, the sooner he could think about math.

Archimedes looked at the sea and the bright sun. He knew that intense heat could be made with sunlight. Perhaps he thought of the metal cone-shaped goblets that used sunshine to light the sacred fires. Finally he said, "Soldiers and mirrors should do it."

The king thought, *Mirrors? In war?* But he knew better than to ask. Archimedes' ideas were odd, but they worked.

Soldiers holding mirrors were positioned to focus the sun's rays on the fleet of ships. Soon the sails burst into flames and the ships sank. Archimedes had saved Syracuse, and the king was pleased.

Galileo was called the Archimedes of his day because of his love for mathematics. He spent his favorite hours as a child building things, such as toy boats and carts. When he became an adult, Galileo built the world's first telescope. He called it a "spyglass" or a "gazer into the distant stars."

What Galileo saw through his telescope thrilled him so much that he wanted to share his discoveries with the whole world. He wrote, "I give thanks to God who has been pleased to make me the first observer of marvelous things unrevealed to bygone ages . . . I have ascertained that the moon is a body similar to the earth . . . [and] the nature of the Milky Way . . . ."

In the seventeenth century, people believed that the sun revolved around the earth, as Aristotle had taught. But Galileo discovered that the opposite was true—the earth actually revolved around the sun! For his radical ideas, Galileo was placed under house arrest.

Athanasius Kircher, a German writer, also lived during the seventeenth century. Testing Archi-

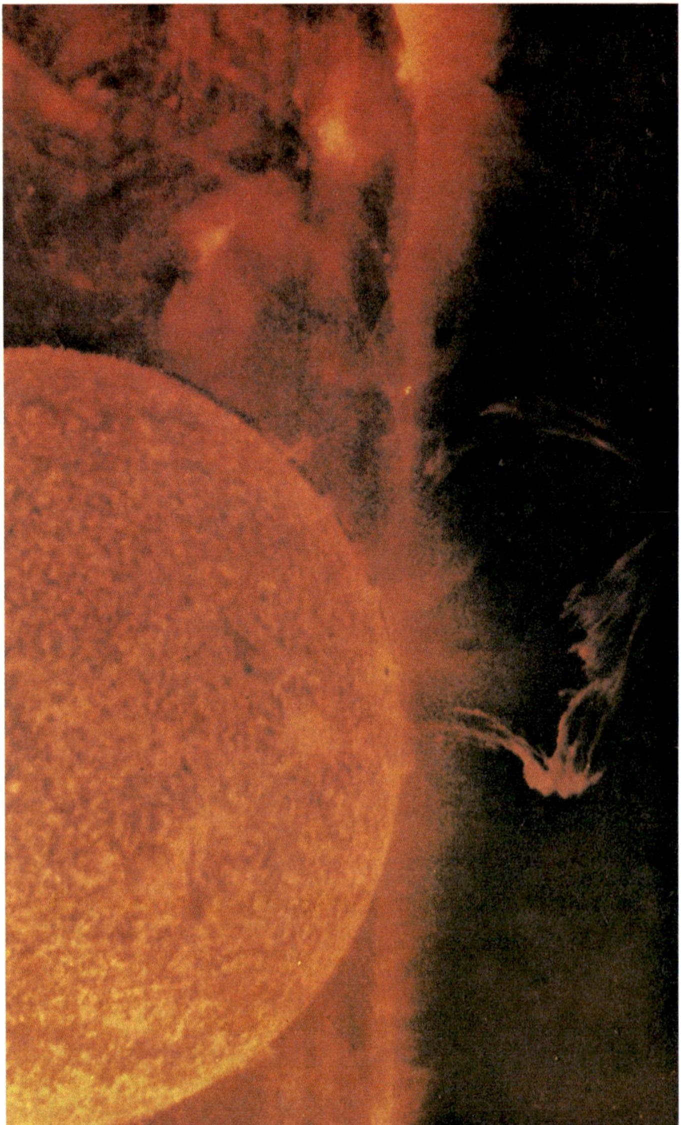

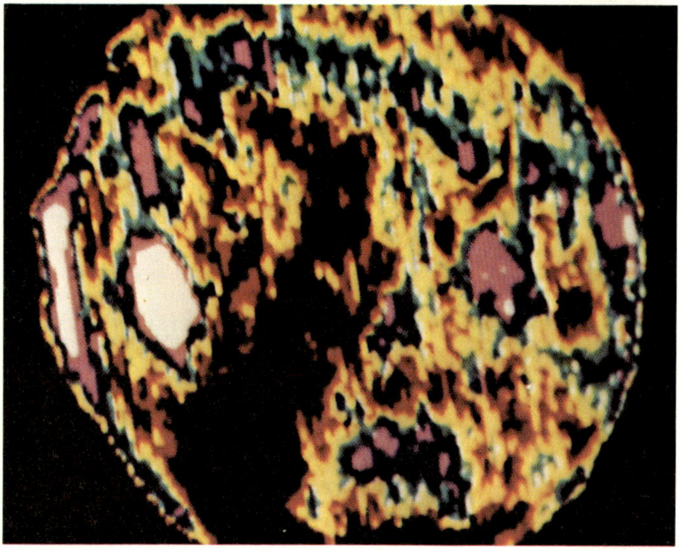

Above—View of the sun's atmosphere from Skylab 3 (Photo JPL/NASA). *Opposite*—Huge solar eruption seen from Skylab 3 (Photo JPL/NASA). *Below*—Photovoltaic array in Schuchuli, Arizona (Photo Solar Energy Research Institute).

medes' theory, Kircher directed the sun's rays to a stack of wood with five burning mirrors. The wood caught fire and blazed.

Two scientists from Florence, Italy, went even further. Targioni and Averani read about how Archimedes had burned ships and Kircher had burned wood with the sun's rays. They decided to try their own experiment. By using glass and sunlight, they melted a diamond.

Antoine Lavoisier, a Frenchman known as the father of chemistry, knew that sunlight could produce very high temperatures. To find out *how* high, he had some glassblowers make a lens of two

glass disks. The main lens was fifty-two inches in diameter. The smaller was eight inches in diameter. Sunlight poured through the curved lens and created so much heat that Lavoisier melted steel and platinum. His solar furnace reached temperatures as high as 3190 degrees Fahrenheit. Lavoisier once said, "The fire of the sun is purer, more potent than a furnace."

The first solar cooker was invented by a Swiss philosopher, Nicholas de Saussure, who put together a set of concentric glass containers (one

inside the other). In his first experiment, de Saussure cooked a bowl of soup. Other types of solar cookers have been developed since that time. One kind resembles a box that can be neatly folded up and packed away. Another resembles a mirrored umbrella with a flat grill at its focal point. Still another cooker is cone-shaped and made of polished metal. Regardless of the shape, these solar cookers use the same principle. They concentrate

Above—Solar collector furnace. (Photo Solar Energy Research Institute)
Top Left—Line-focus parabolic mirror. (Photo Solar Energy Research Institute)
Bottom Left—Power tower, Sandia Labs. (Photo Solar Energy Research Institute)

sunlight to produce high temperatures for the purpose of cooking food.

Augustin Mouchot, a professor of physics, spent twenty years working on different solar experiments. He designed a cone-shaped collector, and exhibited solar furnaces in Paris. One of his projects involved the help of a friend, Abel Pifre. They made a metal reflector in the shape of a shortened cone. Glass tubes were attached to the focal point. The metal cone and glass tubes were attached to a water boiler. When they finished, they had constructed a steam engine. It was used to run a printing press in Paris. The newspaper was called *Le Soleil*, which means *The Sun* in French.

5

HOW DOES SOLAR ENERGY WORK?

When first designing satellites, NASA experts faced a problem. If a spacecraft were to beam down radio messages, it would need electricity. But where would the spacecraft get its electrical power?

They knew that the most available resource in outer space was the light energy of the sun. RCA and Bell Laboratories had already worked with the idea of changing sunlight into electricity. Bell had even made the first solar cell. But it was an unnecessary novelty, because supplies of oil and gas were plentiful during the 1950s. The country did not need more energy at that time.

However, NASA went to Bell. They experimented until possibility became reality. Through their research, satellites and spacecraft were rocketed into orbit with solar cells to power their scientific instruments. From this success, a new energy form was born—solar photovoltaics.

Photo means light in Greek. *Volta* comes from the name of an Italian physicist, Alessandro Volta,

Above—Montezuma's Castle, Arizona. (Photo Solar Energy Research Institute)
Top Right—House with solar hot water panel, around 1900, Orange County, California. (Photo Solar Energy Research Institute)
Bottom Right—Solar dairy barn, New Jersey. (Photo Solar Energy Research Institute)

pioneer in the study of electricity. *Photovoltaics* means the changing of sunlight to electricity. When the sun shines on certain light-sensitive materials, electrons are set in motion. This produces an electrical current.

Not everyone has heard of photovoltaics. The process is not well known because the solar industry is in its infancy. Someday, when solar energy fully contributes to our energy supply, photovoltaics will be a household word.

One way to make electricity directly from sunlight is to use solar cells. These cells make no noise. They are flat, and have no moving parts. Solar cells are reliable and easy to operate. While they can be made of cadmium sulfide, cadmium telluride, or gallium arsenide, the chief material for solar cells is *silicon*. It is the second most abundant element on earth. Sand on the beach is made of silicon

(quartz). However, the sand for solar cells must be refined and extremely pure.

When you place a single grain in a hot chemical bath, the silicon seed grows. The method is very delicate. As the silicon grows, it must be drawn into ribbons or pulled up in sheets. When finished, it looks silvery-smooth. It can be grown in different shapes; as ribbons, sheets, or ingots of varying sizes.

Ribbons can be grown to one hundred feet long. A thin die (a mold that sets the shape) is placed in the molten liquid. As the silicon grows, the ribbon is

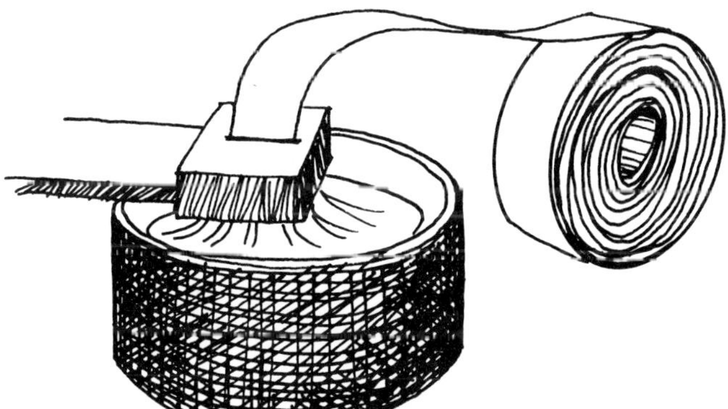

pulled upward through the die, which holds the shape as it hardens. For storage, the mirror-smooth ribbon is rolled onto a drum. It is the cheapest way to make solar cells, as ribbon can be cut with

Above—Solar-powered irrigation. (Drawing Ken Hodges, JPL/NASA)

Top Right—Flat plate collectors, Bozeman, Montana. (Photo Solar Energy Research Institute)

Bottom Right—Solar thermal grain dryer, attached to silo. (Drawing Ken Hodges, JPL/NASA)

scissors and the silicon squares are made without wasteful sawing and polishing. When placed together, the squares fit closely, thereby providing more electricity. (If circles were placed side by side, there would be gaps and less electrical flow.)

Sheets are grown in the same manner. Although the sheets are thin, they cover a large area.

An *ingot* resembles a fat salami, about three feet long. It is extremely hard, even harder than marble to cut. A saw with diamonds embedded in the blade slices the silicon "salami" into wafers twelve one-thousandths of an inch thin. The diameter varies from two to six inches. Part of the carefully grown ingot ends up as costly sawdust on the floor.

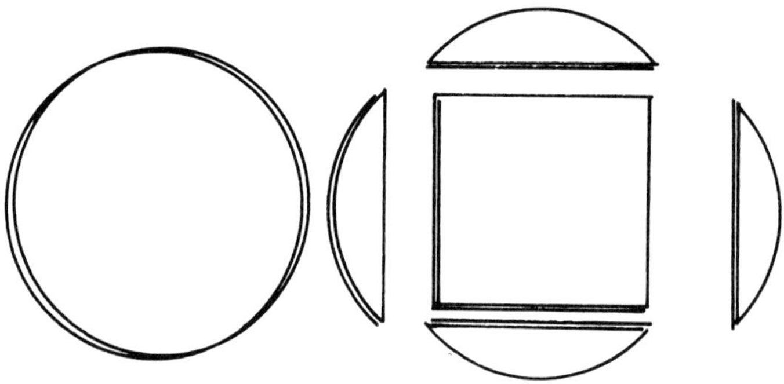

More silicon is wasted if the round slices are cut into squares. (When a square is cut from a circle, the four rounded edges are wasted.) Another expense arises when the saws are replaced, as they are only good for so many cuts. Such waste makes solar cells more costly to produce.

The silicon squares or wafers are then etched in acid. A mesh of thin wires connects two wafers sandwiched together to form solar cells. The solar cells are often colored a lustrous blue. The cells are arranged on a flat panel made of glass, sheet metal, or wood, and sealed with glass or plastic.

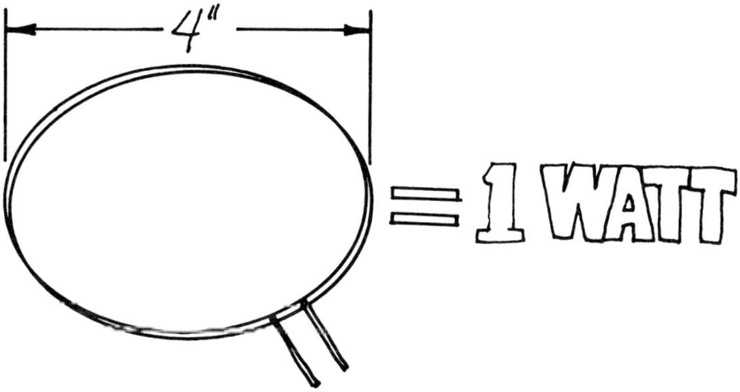

One round solar cell, four inches in diameter, produces one watt of power. To power a one-hundred-watt light bulb, you would need one hundred solar cells.

Solar experts measure the concentration of light on solar cells in "suns." A five hundred power concentration is referred to as "500 suns." A one thousand power concentration is called "1000 suns."

Solar cells are like building blocks. When connected electrically, they form a power system called a *solar module.* Solar modules may be combined into bigger *solar arrays.* These involve so much voltage that a regulator is needed to control the electricity and to direct it to a storage unit.

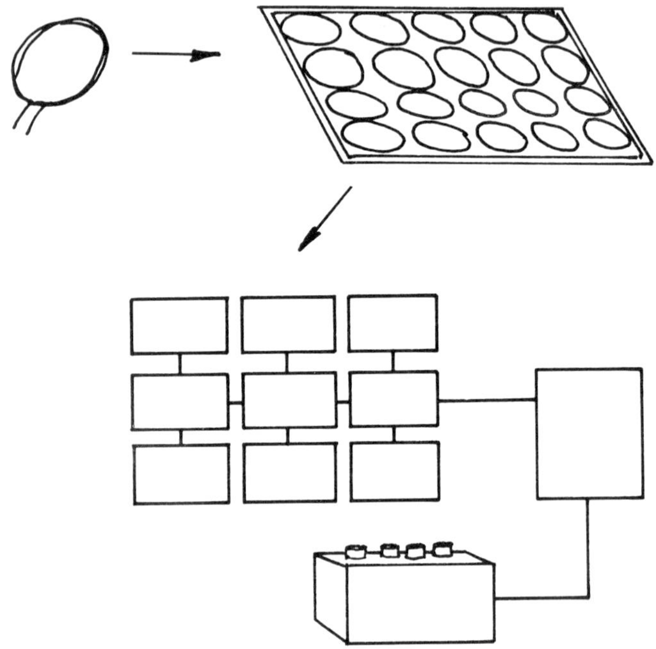

To capture the most sunlight, you need to know how the sun's angle differs with the day and seasons, the total number of sun hours, and weather conditions, since the thickness of clouds affects the amount of sunlight reaching the solar array. Arrays must be tough, to withstand the attack of any kind of weather. Sometimes, however, problems arise. Solar cells can crack. Dirt can cause a power loss. Or, if the solar cells become overheated, the metal connectors may melt. But these incidents are rare. Solar cells are designed to last approximately twenty years. They were conceived for use in space, with quality and compactness in mind.

Of all the solar arrays ever built, the largest was for the Skylab space station. Ten thousand watts powered Skylab and kept its instruments working. It was America's first manned vehicle flown in space with solar panels.

For earthly purposes, solar arrays are used in different ways. Solar cells may provide power for wrist watches, calculators, radios, or walkie-talkies. Solar arrays may be placed directly on homes or buildings, with an electrical generator nearby.

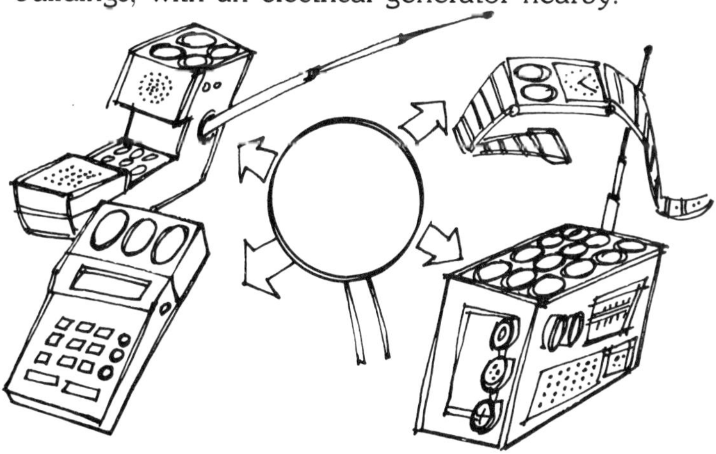

Mirrors Reflecting Light

Another way to make electricity from sunlight involves using mirrors made of spun aluminum, glass, or other materials with a silvery surface. It is called the *solar conversion method.* This process transfers energy from one form to another: light, to heat, to electricity.

Mirrors reflect and concentrate sunlight so that high temperatures produce steam for making electricity. Five basic steps are involved:

1—Mirrors reflect the sun's rays toward

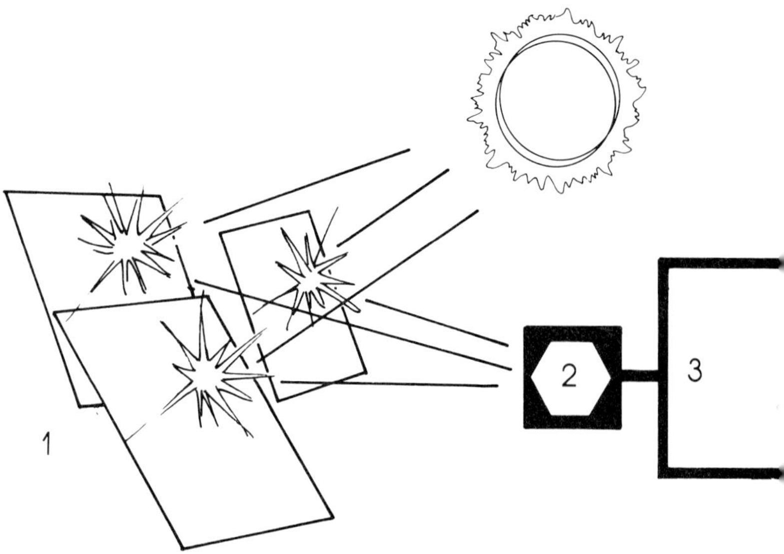

2—A receiver, such as a tower or pipes, which absorbs the sun's energy.

3—Fluids transfer the heat to storage or to the generator.

4—A tank stores the heat for use when the sun does not shine, such as during nights or cloudy days.

5—A turbo-generator makes electricity.

In these five steps, the sun's rays are collected with mirrors, absorbed as heat by either a *power tower* or *pipes*, and used to make steam, which spins the turbines to make electricity.

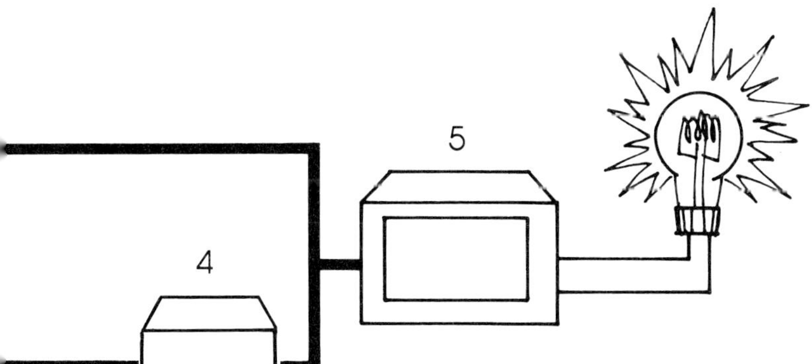

If you wanted to build your own power tower, here is what you would do. Take a large area of land. Put up a very tall tower with a built-in water boiler in the center. Surround your tower with hundreds of heliostats, or huge curved mirrors, circling outward for a mile. Make sure your mirrors are lightweight. You want the mirrors to be light so they can track the sun's movements. Then mount the motor-controlled mirrors to follow the sun in east-west, north-south directions.

With all the mirrors facing the power tower, they will catch the sunlight and beam it to the power tower to one precise spot. Once inside the tower, the light is trapped. The concentration of sunlight causes intense heat. Temperatures sometimes reach one thousand degrees. The high temperatures generate so much heat that the water in the boiler changes to steam. The steam moves the turbine, which is hooked to a generator. The generator then makes electricity.

While these "mirrored systems" are clean, they can create strong winds. The heat in the tower causes constant uplifts of air. They also take up a lot of space on the ground.

As a rule, solar power plants should be near the buildings they supply with electricity. There is loss in electrical power when the energy is transported over great distances. Energy from the sun is free.

However, the costs for land, equipment, and power-tower installation are high.

The Italians built a power tower in St. Illario-Nervi. Two hundred and seventy mirrors make up the collector array. It can produce 150 kilograms per hour of hot steam at five hundred degrees Centigrade.

America's first power tower is a ten-megawatt electrical plant on 130 acres in Daggett, California. This pilot project cost $140 million. It is operated jointly by the Department of Energy (DOE), the California Energy Commission, Southern California Edison, and the Los Angeles Department of Water and Power. Solar One, as the project is known, is the first of many power towers in the United States.

There is still another solar thermal method, one that uses *flat plate collectors.* Like power towers, flat plate collectors take up vast stretches of land. The mirrored collectors are arranged in rows, one after the other, for as many miles as needed. The flat plates tilt sunward, automatically following the sun's path during the day and different seasons.

Flat plate collectors direct their sunlight either to a central receiver, such as a tower, or to pipes located beneath the mirrors, which carry the heat to the power plant. The heat then creates the steam needed for spinning the turbo-generators.

The newest *point focus* technology was developed by the Jet Propulsion Laboratory in Pasadena, California. Point focus technology applies the same light-catching principle to convert sun heat to power. Mirrored reflectors are dish-shaped (parabolic), and curve to concentrate sunlight at the focal point of the collector. Each mirrored reflector has its own receiver, heat engine, and generator at the focal point. Each dish is an independent generator.

A typical unit ranges from thirty to forty feet in diameter. The receiver is the size of an oil drum. Each generator would deliver fifteen to thirty kilowatts of electricity. Many of these mirrored point focal reflectors would make up a power plant.

From Greenhouses to Heating Houses and Water

Do you have a greenhouse in your back yard? For years, gardeners have used greenhouses to grow their plants. Greenhouses, mainly made of glass, let the sun in and trap most of the heat.

In the same way, solar collectors capture the sun's energy. Sunshine passes through the glass and is absorbed by black material. When this happens the sun's energy cannot escape. It can be used for heating water, or for heating houses. Pipes take the hot water or hot air through the house. A storage system provides heat when the sun is not shining.

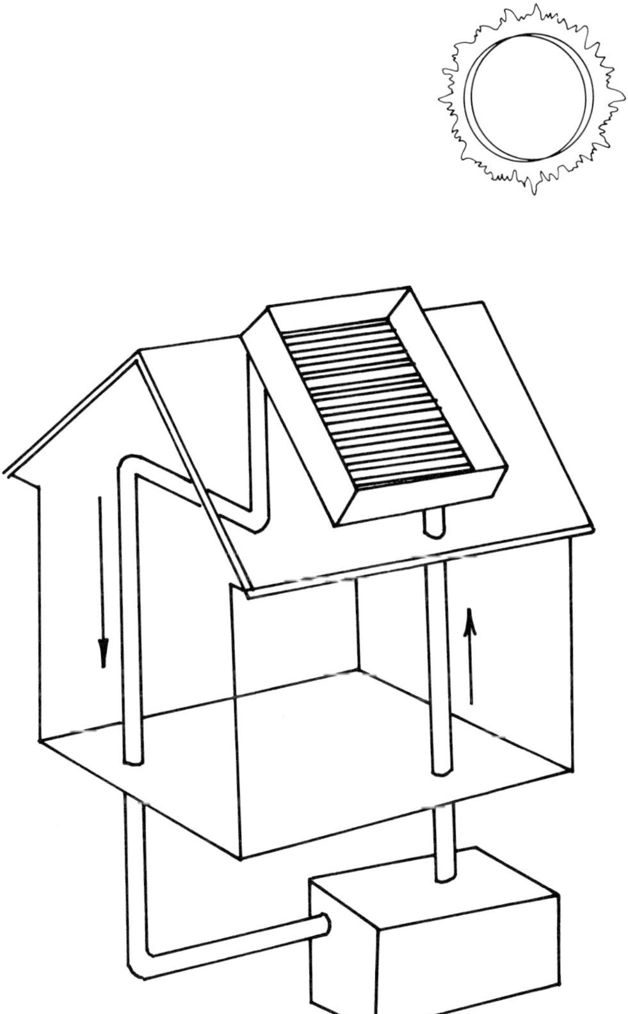

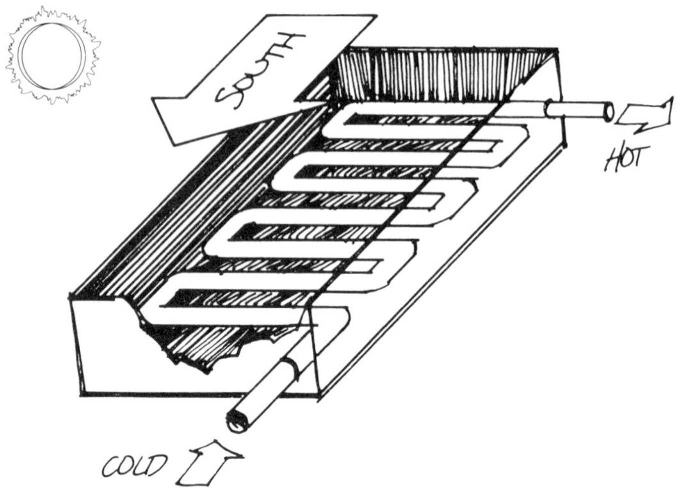

What does a solar collector look like? The bottom of the rectangular box is black, because black absorbs heat. Coils and pipes are attached. Through these, water or air may circulate. The collector is covered with glass to protect the unit from weather changes. Then the solar collector is placed on a sun-exposed roof that faces south, so it will receive maximum sunlight.

Black material on the bottom of the collector increases the heat. This heats the air or water, which moves from the roof through pipes to a system that heats the house or building, or makes hot water. Although most solar collectors are similar, there are different systems on the market for distributing and storing the heat or hot water.

Cool Air from Sun Heat

Sunshine may be used for air conditioning, too. Solar cooling requires refrigeration equipment that uses solar heat instead of natural gas. The design is complex, but results prove that some solar air-conditioners offer money-saving advantages. We need cool air the most during daytime in the summer when the sun sizzles at its peak.

Setting up solar heating and cooling systems for homes and factories is expensive. However, once the system is paid for, the warm or cool air and hot water are free.

6

WHAT IS HAPPENING TODAY?

Everyone needs water, not only to drink, but for food and other products. It takes eight thousand tons of water to grow one ton of wheat. Thirty-one thousand tons of water are needed to raise one ton of beef!

In some parts of the world, where the only water is murky or from the sea, the sun offers the only hope of getting fresh water. You can get as much clean water as you want from a simple process called *distillation.* You need a solar still or tank, covered with a slanted sheet of glass. The sun warms the water and causes it to evaporate. The water vapor cools as it rises onto the glass above. The pure vapor condenses into drops of drinkable water that slide down the glass into a bin. And there you have fresh water!

One of the first solar stills was built in Las Salinas, Chile, in 1874. The water of Las Salinas was salty. The still provided six thousand gallons of water daily for the miners and donkeys for thirty years, until pipes brought water to the desert.

Farms everywhere need water for growing crops. Usually irrigation systems require pumps that run on power from a utility company. But the Dalton Cole farm near Coolidge, Arizona, has installed a 150-kilowatt solar thermal electric power system. They use trough-shaped mirror concentrators to focus sunlight on pipes to generate heat. The heat is then converted to electricity, which powers the irrigation pumps. Fourteen hundred gallons of water are provided for Dalton's cotton crops each minute.

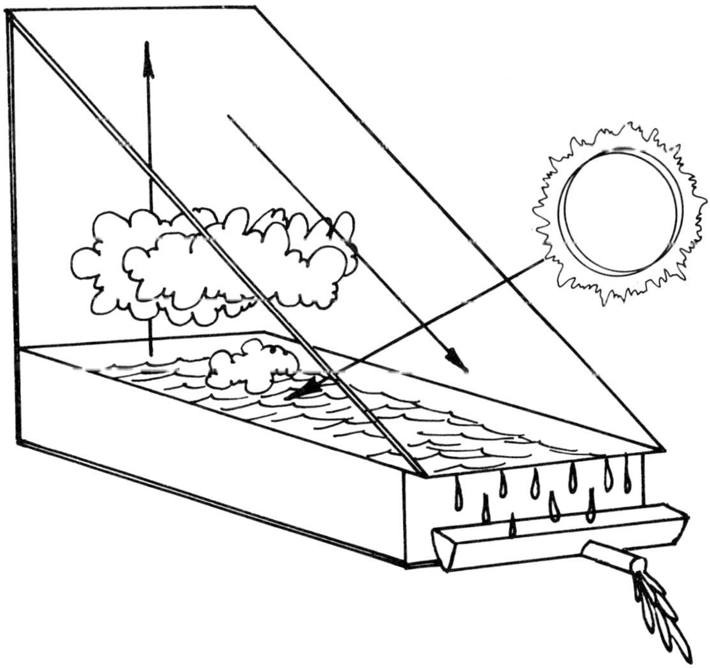

For centuries, farmers have dried their grains in the sun. If crops such as corn are harvested with lots of moisture, they must be dried before being stored. Otherwise, they will rot.

Many farmers have attached flat plate collectors to their roofs to dry their grains with large amounts of solar-heated air. This is a slow process that requires low temperatures. Farmers also use the heat for drying produce. All kinds of spices, fruit, and vegetables may be sun-dried. Onions, parsley, mint, raisins, apples, and dates are a few examples. Sun-drying not only preserves food, it enhances the flavor as well.

One Illinois farmer had a huge hog house built with solar collectors lining the entire roof. It worked

so well that he and two friends started their own company so they could help other farmers solar-heat their hog houses and chicken coops.

One of the poorest areas of Los Angeles County is the Antelope Valley. Federal funds were used to convert the area into "Sun Village" to help the families reduce the growing cost of electricity. Seventy-five homes were outfitted with solar panels to provide heat at low cost. There are nine solar energy "demonstration homes" in Sun Village, and more are to be built and sold at low interest rates.

Many schools are using solar energy. Timonium Elementary School in northern Maryland was one of the first schools to air-condition and heat one of its wings with solar energy. Glass panels on the roof catch the sun's rays. Sunshine heats the water, which circulates through a fifteen-thousand-gallon storage tank, which is used for warming or cooling the air.

Martha's Vineyard High School in Massuchusetts needed more space. But the school board wanted to conserve energy. They designed an energy-saving addition and built the classrooms so that windows faced the sun. Daylight provided most of the lighting inside the classrooms, workshops, and kitchen. Double-glazed window edges prevented much of the heat from escaping. Now the school uses twenty percent less fuel oil, even though the school is one-third larger.

As a science project, students at Crescenta Valley High School in La Crescenta, California, built Sunfire, a solar generator that stands four stories high. The parabolic (dish-shaped) trough catches sunlight with 240 handmade mirrors. The sun's heat is reflected onto a water chamber. The heat turns the water into steam. The steam turns a turbine that generates five kilowatts of electricity. It took five years for successive classes of students to finish the project.

7

WHERE IN THE WORLD?

America has its sights set on the sun. Many hope that, by the year 2000, fifteen to twenty percent of all the energy we use will be solar. But solar cells are still too expensive to mass-produce.

Cost was not a problem for our space program. The solar cells used on satellites were handcrafted with the precision of a Swiss watchmaker. They

required painstaking time and effort. They cost a fortune. To be commercial, solar cells must be easy to make and cheap to buy. They must compete with other forms of energy in price. Otherwise, they are not worth developing.

The promise of solar cells has not been fulfilled because they are so difficult to manufacture. For all the work that has gone into photovoltaics, we still need a breakthrough.

The United States is pouring millions of dollars into research and trial projects so that problems in solar cell technology can be ironed out. Efforts are backed by the Department of Energy. These projects have brought scientists, industry, universities, and even foreign countries together. The exchange of ideas will push our progress forward, until we develop solar cells that are cheap and easy to use.

When you trace what has happened since our space program, you can actually see the solar field growing. Early uses for solar arrays were modest. They powered a highway emergency call box in a national park; radio equipment for the U.S. Weather Department; a railroad crossing signal in Rex, Georgia; and a U. S. Coast Guard buoy in Long Island Sound.

Later experiments were larger. An entire Indian village in Schuchuli, Arizona became solar-powered. Lights, refrigerators, washing machines, water pumps, and other appliances run on a 3.5-kilowatt array. For the past few years, an eighty-acre cornfield near Mead, Nebraska, has been watered with a solar irrigation system. The Mount Laguna Air Force Station in California uses a sixty-kilowatt array. The Mississippi County Community College installed a three-hundred-kilowatt photovoltaic system in Arkansas.

Factories have now begun using solar energy to meet various needs. Campbell Soup uses it for washing cans. Tropicana uses it for pasteurizing orange juice. Gilroy Foods uses it for drying onions. Anheuser Busch uses it for brewing beer. Riegel Textile Corporation uses it for dyeing fabrics.

Some of the nation's largest corporations have entered the solar field. General Electric, Exxon, Texas Instruments, Westinghouse, Arco, and Shell Oil are only a few.

The federal government offers a tax credit to homeowners with solar systems. Some states offer tax cuts, too. Thousands of homes were heated or cooled by the sun last year. The exact number of installations in the United States is unknown. One fact stands out, however. Increased sales in solar units prove that people at home and in industry are using more and more solar energy.

France

France has a long history of using solar energy. Its government has always believed in sun power for two reasons. First, sunshine is renewable. Second, the French do not want to depend on other nations for energy.

Once France had a sufficient supply of coal. Coal was the basis of its industrial strength. Now, the nonrenewable supplies of coal are at an all-time low. Sixty percent of the coal being used today is imported from Poland and West Germany. To pull away from foreign dependence, France encourages its citizens to use solar energy, and offers them tax benefits if they do.

Located near a ski resort in the Pyrenees Mountains is the world's largest solar furnace. The Odeillo Furnace towers 133 feet high and is 165 feet wide. The main mirror stands eight stories tall!

Facing the main mirror are rows of smaller flat mirrors that follow the sun as it moves through the sky. The concentrated light beams on target and causes heat. The Odeillo furnace produces temperatures hot enough to melt a thick steel plate in three and a half seconds.

Brazil

Brazil, the largest country in South America, has all kinds of solar projects underway. Recently a four-hundred-bed hospital changed to solar energy. One thousand flat-plate collectors were installed to provide steam and hot water. Before the installation, the hospital used eighteen hundred tons of fuel oil every year. With its solar panels, the hospital saves fifteen percent on energy.

University researchers contribute their ideas, too. One campus built a large ice-maker for preserving food with solar refrigeration. Another university built a solar seed-dryer for Latin American farmers. At another college, solar heat engines are being designed for local manufacturers.

Recently the Brazilians spent a million dollars to build a solar power tower, the first in Brazil. Solar experts hope to spark the interest of local industry with this project. The power tower will provide one hundred kilowatts of electricity.

Israel

A solar pond is a large body of water, natural or man-made, that has been converted to produce electric power. Dark, heavy rubber lines the bottom. A dike encloses it all around. If the pond is man-made, then the waters are "doped" with raw salts. This is not necessary if the water is naturally salty, as in the Great Salt Lake of Utah or the Dead Sea of Israel.

Sunlight shines on the water, and heats the salts at the bottom of the pond. The salts dissolve. The melted salt thickens the bottom layer and creates a temperature of two hundred degrees Fahrenheit. Meanwhile, the upper layer of water separates itself

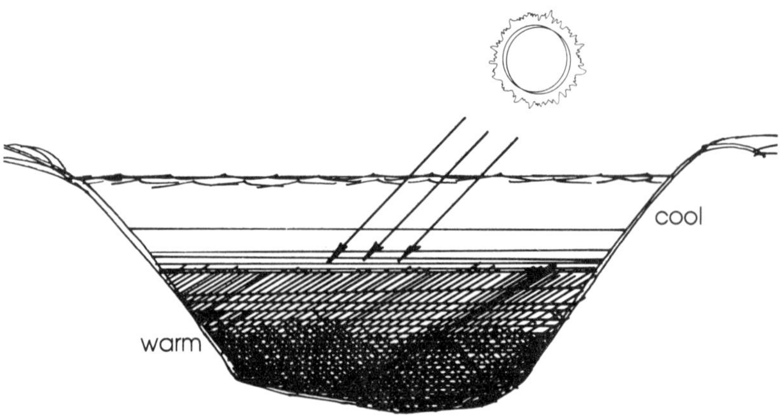

from the salty solution. It remains cool and clear. The sun's radiant energy becomes trapped by the salts below, but does not mix into the water above. Thus, a solar pond contains two layers: cool and clear on top, dense and hot on the bottom.

Israel opened the world's largest solar pond at Ein-Bokek on the Dead Sea. The Ein-Bokek power station was a pilot program. It proved that solar energy could be generated day and night, winter and summer, whether the sky was cloudy or not. Its success has led to tremendous plans for extending solar ponds in Israel.

Israel's Ministry of Energy is building a five-thousand-kilowatt power station in Ormat. It has a total capacity from the Dead Sea for two thousand megawatts. Israel feels this technology marks the beginning of the end of their oil dependence on OPEC.

Australia

Three hundred hours of sunlight shine on Australia each month. Its closeness to the equator and the sun makes Australia a natural for using solar energy, and there are many solar companies located there. They manufacture everything: hot water heaters, swimming pool covers, solar cells, cooling systems, appliances, and solar hardware.

These products, exported abroad, are noted for their fine quality.

Australia has a program that welcomes visitors from other countries to see how well their solar hot water systems work. Part of the tour includes the Somers Yacht Club near Melbourne, the Sacred Heart College in Adelaide, and a factory where cans are warmed in New South Wales. They are developing solar-powered air conditioners for Australia's mining areas.

Solar hot water heaters are widely used. In some parts of Australia they are required by law.

Mali

It is surprising to find solar technology in Mali, a West African nation near the southern rim of the Sahara Desert. Only ten percent of the natives are educated. Their culture is rich, but the nation is poor. How is it then that Mali has been using solar energy for twenty-five years?

One reason for this progress is outside help, which started with France and its interest in solar energy. Years ago, Mali was a French West African colony. When Mali gained independence, its people wanted prosperity. But they suffered a terrible drought that affected livestock and crops.

The Mali government welcomed foreign firms to

invest. It offered a liberal payment plan that attracted investors. The climate provided a lot of sun and little rain. The United States offered aid and a solar development program that created jobs for the people.

Sofretes, a corporation from France, has built four solar irrigation systems in Mali. The European Development Fund has ambitious plans for building a major solar power plant.

Solar energy is being used for air conditioning, crop and fish drying, hot water systems, and water purification systems that produce drinkable water. Mali was the first country in Africa to install photovoltaic water pumps. In Dire, a town on the banks of the Niger, you could see an eighty-kilowatt power plant being built.

Other countries in Africa have had solar systems brought to them: Kenya, Nigeria, Senegal, Cameroon, and the Ivory Coast.

8

WHAT ABOUT THE FUTURE?

Wherever you turn today, you can hear about sunlight and solar cells powering toys, refrigerators, airplanes, cars, and railroads. Practical experiments are expanding as our imaginations expand. Ideas are booming. But the sun still remains a mystery.

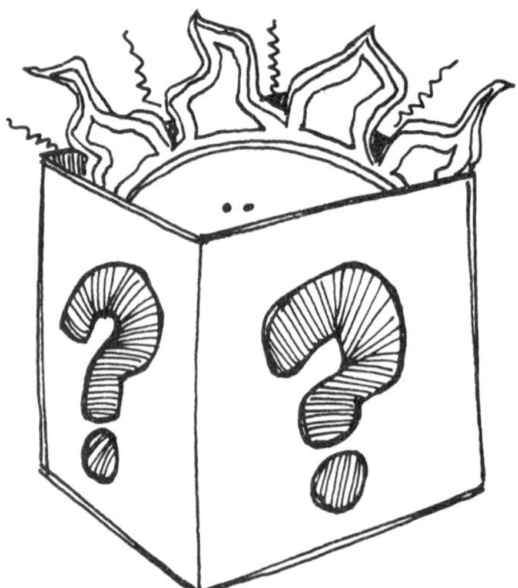

What are those webs of magnetic fields on the sun? What causes solar winds to change and affect earth's upper atmosphere? Is there any relationship between the intense cold felt on earth between the years 1650 and 1710, and the disappearance of sunspots during this time? Can we blame cold climate on low sunspot activity? These and other questions perplex the experts.

One obstacle limiting our scientists is that the earth orbits the sun on one plane. The telescope view reveals the sun's equator because we circle the sun at that level. No spacecraft has ever gone more than a few degrees above or below that line.

Scientists want to study the sun from different angles, such as its north and south poles. To accomplish this, a unique space launch called *Solar Polar Mission* has been proposed.*

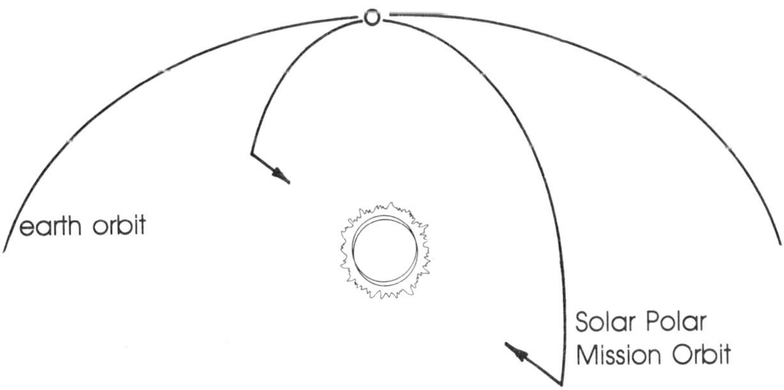

*As of this writing, the Solar Polar Mission has been suspended indefinitely.

The United States' National Aeronautics and Space Administration (NASA) and the European Space Agency (ESA) have agreed to send two spacecraft to study the sun's never-before-seen polar regions. The European Space Agency represents these countries: France, Italy, Holland, West Germany, Switzerland, Spain, Great Britain, Sweden, Belgium, Denmark, and Ireland.

Rockets that will move the spacecraft out of earth's orbit will be fastened beneath the Solar Polar satellites. Then, rockets and satellites will be put inside a space shuttle, which will be manned by three astronauts.

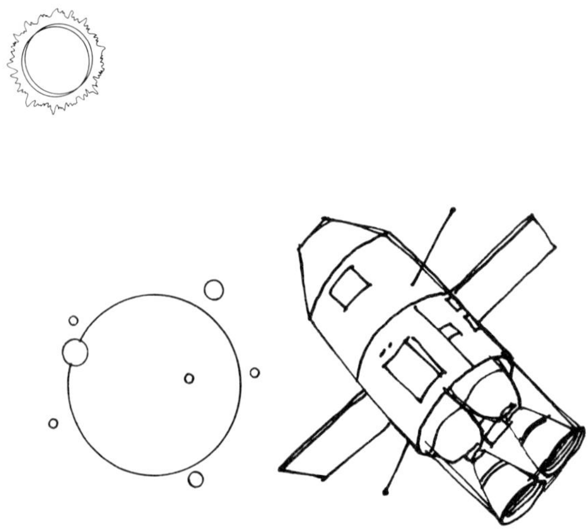

Once the space shuttle reaches earth's orbit, the astronauts will open the cargo-bay doors. Mechanical arms will lift the Solar Polar spacecraft outside, into space. Then the astronauts will move to a safe distance away from Solar Polar. Using their radio signal, they will fire the rocket booster engine. This will kick Solar Polar out of earth's orbit and send it off to Jupiter.

The journey to Jupiter will take between 470 and 475 days. The tremendous gravity there will thrust the satellites toward the sun. The planet Jupiter's gravity will be used as a slingshot.

One of the satellites will swing around the sun's

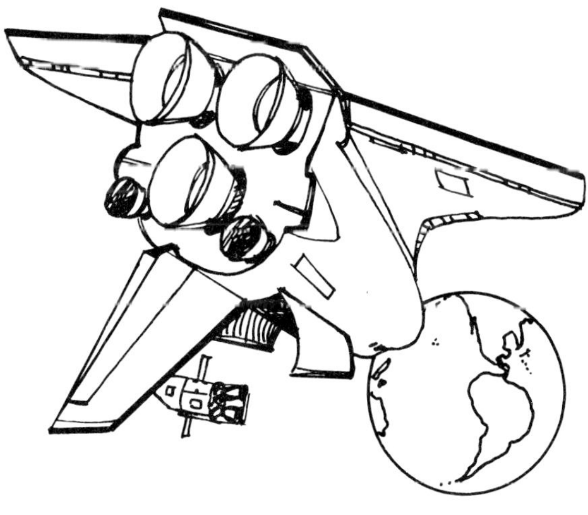

poles from north to south. The other will cross the poles from south to north. Each spacecraft will spend nine months going from pole to pole.

The satellites will not take pictures with cameras. Instead, two instruments will make detailed images of the sun, as well as analyze and feed information about cosmic dust, solar winds, and other things into a computer.

What we learn will be invaluable. The knowledge we gain will make outer space less foreign. Someday we may regard outer space as an extension of earth's back yard.

Men and women continue to confront current challenges, dream new ideas, and find solutions to meeting our energy needs. Scientists all around the world are seeking answers; and many ideas are emerging.

Solar energy will, through our creative thinking, become a strong force for the future.

BIBLIOGRAPHY

1. Behrman, Daniel. *Solar Energy: The Awakening Science.* Boston: Little, Brown, 1976.
2. Branley, Franklyn M. *Solar Energy.* New York: Thomas Y. Crowell, 1957.
3. Freeman, W. H. "The Solar System." *Scientific American*, 1975.
4. Gadler, Steve J., and Adamson, Wendy Wriston. *Sun Power.* Minneapolis: Lerner Publications Co., 1978.
5. Halacy, D. S., Jr. *Coming Age of Solar Energy.* New York: Harper & Row, 1973.
6. Hoke, John. *Solar Energy.* rev. ed. New York: Franklin Watts, 1978.
7. Knight, David C. *Harnessing the Sun.* New York: William Morrow & Co., 1976.
8. Metros, Thomas H., and Bitter, Gary G. *Exploring with Solar Energy.* New York: Julian Messner, 1979.
9. Rau, Hans. *Solar Energy.* rev. ed. Edited by D. J. Duffin, translated by Maxim Schur. New York: The Macmillan Co., 1964.

10. Schwartz, Marie Sokol. *Harvesting the Sun's Energy*. Fullerton, Calif.: Designs III Printing, 1975.
11. Wohlrape, Raymond A. *Exploring Solar Energy*. New York: World Publishing Co., 1966.

333.8 Yates, Madeleine
YAT
2843 Sun power

DATE	BORROWER'S NAME	
APR		